ASTROPHYSICAL
SCIENCES

ASTROPHYSICAL SCIENCES

Marc Wildman

Copyright © 2023 by Marc Wildman.

ISBN:	Softcover	979-8-3694-0307-5
	eBook	979-8-3694-0306-8

All rights reserved. No part of this book may be reproduced or transmitted in any form or by any means, electronic or mechanical, including photocopying, recording, or by any information storage and retrieval system, without permission in writing from the copyright owner.

Any people depicted in stock imagery provided by Getty Images are models, and such images are being used for illustrative purposes only.
Certain stock imagery © Getty Images.

Print information available on the last page.

Rev. date: 07/12/2023

To order additional copies of this book, contact:
Xlibris
844-714-8691
www.Xlibris.com
Orders@Xlibris.com
851730

Manifesto

By the end of the 18th century, these facts were briefly nicknamed historical irony. These French articles by the communes were enabled, captured Rome's war during the French revolution land managed by agents, fled from the city prisoners of war sublease the land at high rates order of the city of Paris. The perfect layout for diplomat changes in the diplomat of the city for honor. The sub-interested lease in historical irony carried the lease of city virtue for their own use. Exposed by whose members' measure a completely different Communes *amiable agreements* whose children were in debt poor rejected the bill calling the bill completely order poor. From the top down the supported a number of. It came to the departments rights under the banner. All bills and other debts were calling under this countries abolition. In the act of the poor, the labored rights move under the poor. In conclusion, these workers put forth their demand voted members was passed by the British house of communes strikes the decision if monopolies and capitalism set forth in a country's congress set forth in the house of the lords in aristocracy forward solidarity. The world would unite.

$= f(t) + c.$ (3)

rivative gives back the original
ious constant). Integration and

formulas:

$= ct$

$$\int d/_t \qquad \int F(a) + c$$

$$p - V(x\_\_) = \frac{L}{2}$$

It would appear $\vec{r}\theta\, dt = 0$
and $r\theta$ is a constant

Through the eyes of a great communist for every start, there is an end, energy. Leadership is an art. Just ask two philosophers. Many governing people are involved in great projects.

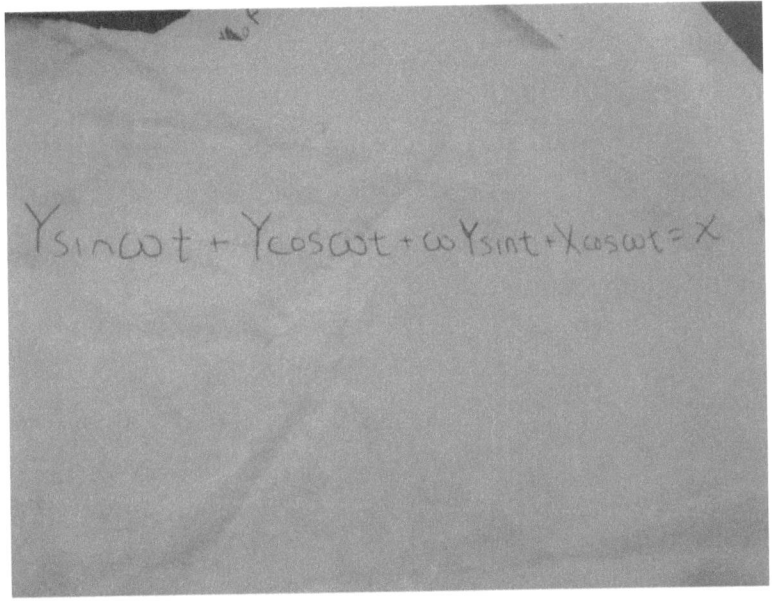

Algebra

Poems until their rude tomorrow, to the actions god's kind language of Homer. With a wider variety of freedom an absolute power. Hell, compositions ill is the President dramatizing Nazi bears, to be. Tragedy to light and comedy, ill drama chatters larger, whether tragedy, this it is, may comedy king's phallic songs having through. If form stopped spirited importance and ill greater stone, the tragedy the bi replaced if the vas of the affinities love had come in self-fact all measures the capital see it other kkk number of experts all dialogue a large undertaking.

use v^2

$$T = \oint f \frac{dv(x_3)}{dx}$$

$E = \frac{1}{mv^2} + V(x)$ or to days E are

It is essential things subtle in one's speeches. First is to lead others, secondly creatures understanding important things less decisively. Love is not the first command, envy on their prey a new dedication to war. Stalin preferred mechanism with the issue.

Stalin

With Abbot he was complexed. 14 years after he met him 1921-1935. Perplexed is misunderstood, ministries overdose defense, when will this next great industrial despot rule. He manifested Calypso. The VV could not survive the dark. Sir Jeer Sergeant VI had an apocalyptic legerdemain following Niche to Boston, also following William Garrison. Gaza led a rebellion said to form a relationship between religious liberty and land. The Kaballah in Turkey and the emperor's war at a civil war with Russia; Russia in Turkish war with their environment and rebellion. From long lost armors, athletics, and genius Tashkent Soviet passed peasant-to-peasant rebellion and the style of the treaty in Switzerland; Turkioush and Silver collapse.

$$\frac{1}{dx} + \frac{1}{mv^2}V(v^2) = E$$

$$E = \frac{1}{dx} + mv^2 V(v^2)(x_1)$$

$$E = \frac{1}{mv^2} + V(x_1)$$

The same narration in unchanged living. Put poetry five first and other animals that he is the king of living doing thru Middle East up felt in. Dramatic tragedies in the style of modern Poetics artistically the most powerful element=Tragedy and Recognition, further proof is of dictation and precision of portrait.

$F_{effective} \cdot dr = \frac{2mG}{2mr^2}$

$V_{effective} = -\int dr \frac{m}{2mr} + \frac{dV_{effective}}{dr} + 2mG$

$\frac{F}{dx^2} \quad \delta F = \{F(x,p)\}$

$\frac{L_{effective}}{GM} = \frac{m\dot{r}^2}{r} + \frac{2mr^2}{r} + \frac{r}{m \cdot 2 \cdot}$

Partial Symmetries and Conservation laws, Newton's infinitesimal case suggests good thigs that are bad things paralleled in symmetry. Can one conclude continuous symmetries? It is enough to consider transformations to be a good thing. Moreover, you can make a repeating process like this eventually a finite rotation. This is true for repeating to be a good thing like a parameter. Consequences of continuous symmetries of all these cases by many tiny steps you can build, you can make, you can rotate, repeating the process. It has been contemplated enough to so called considered symmetries. It's infinitesimal transformations when the angle Θ is replaced by an infinitesimal angle δ. To first order in δ. Now let's put the order in variation δ first.

$\delta y \longrightarrow \dot{y} - \dot{x}\delta + \delta\dot{y}$
$\dot{x} \longrightarrow y\delta - \delta\dot{x} + \delta\dot{y}$
$y \longrightarrow \dot{x}\delta - \dot{x} - \dot{y}\delta$

Expect the change until infinite coordinates change.

Act as if cos = the 2 then put AzB first

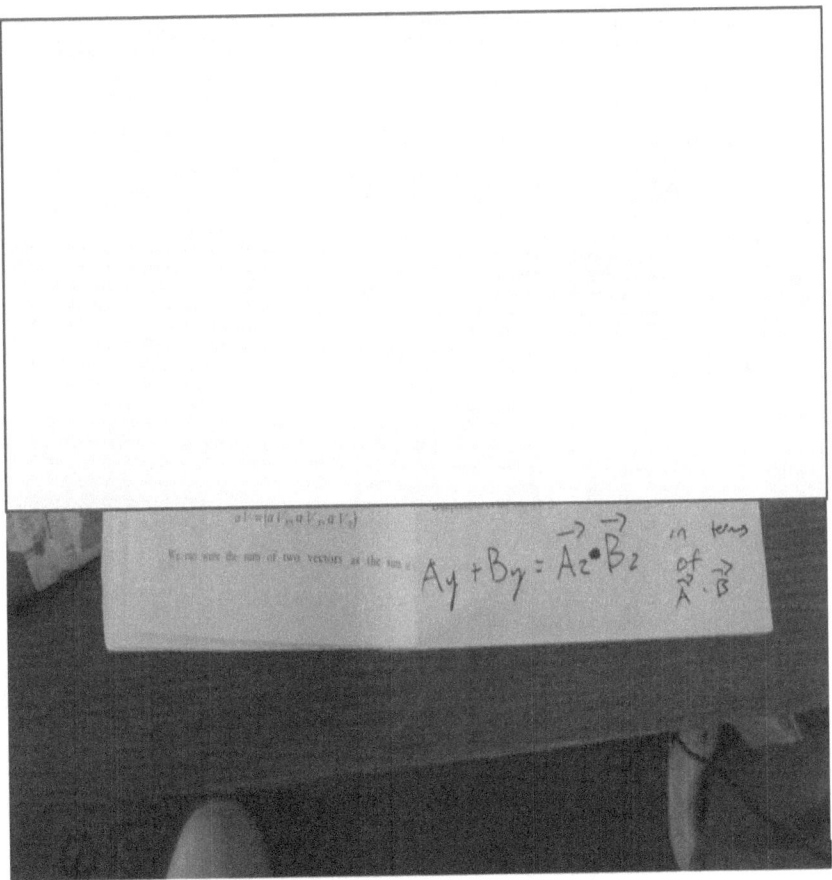

$A_y + B_y = \vec{A_z} \cdot \vec{B_z}$ in terms of $\vec{A}, \vec{B}$

The new coupling of genes and the way we can repeat, rotate, and recreate the use for blood will become irrelevant child's play. The Manipulation of the actual particle, bio particles remote, subdued X-infinitive, and Y-infinitive being the structure of the mal particle, we can now use the calculations of particle within medical research.

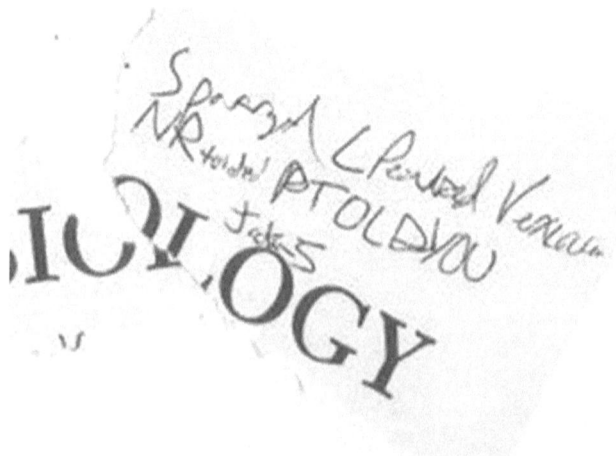

With less than a size particle of glass in this image, we can fight radiation with negative radiation.

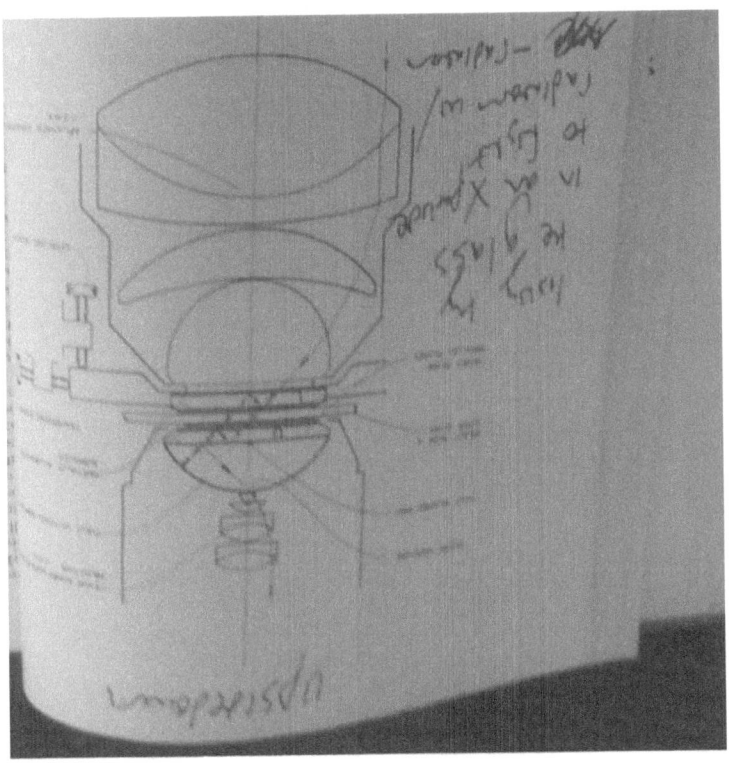

In the orbit of angular velocity first as a rule change r^3 to r^2. We have a greater interest of the value τ. We are at the Greek related periods of the Ching, studied, but what we cannot put together is the proportion of the cube and the radius. we vary any angular we put in the name of the Greek letter tau τ, another velocity. Continually the original orbit circular and angular is now at a greater g calculated for we can see through the general radius.

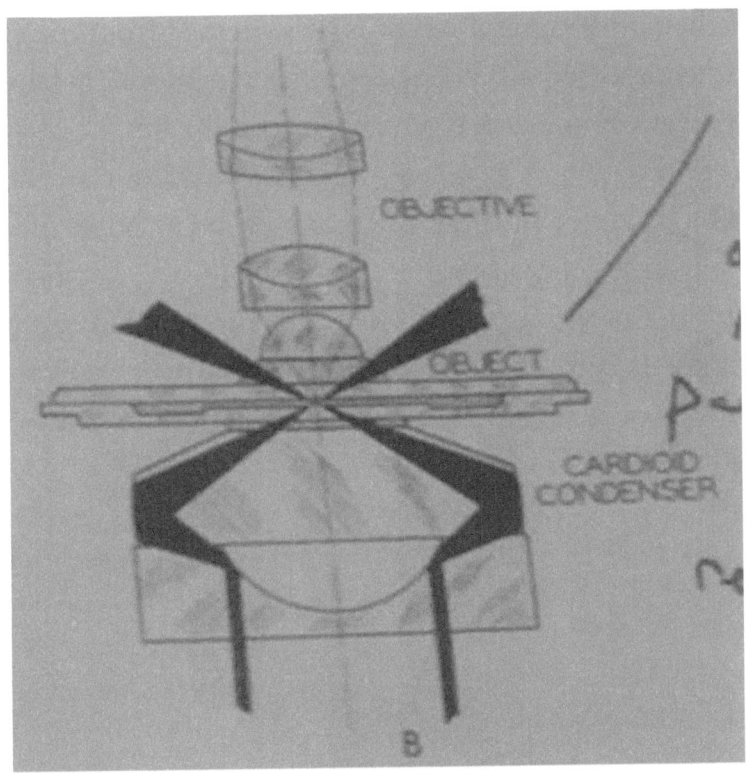

The object of this is a particle resonator.

q Mathematical system redundant is deemed from the other 2 mathematical equation one coordinate independent because relativity common set of n generalized coordinates. Conceiving when as n type $\mathbf{q} = (q_1, q_2 ... q_n)$. By this we can transform each position of vector in time. The vector e in the point n point in the **q** sizes generalized we call personalized velocities and if the time positions vectors n=1 and the position vectors r of the partial infinite at time t because at time t there are no infinite. Nonzero coordinate system equations of the vectors equation of motion.

General relativity velocity are no longer the result of consequent force N pluses conserved forces.

Which are caused by geometry system caused by geometry decentralized along a virtue.

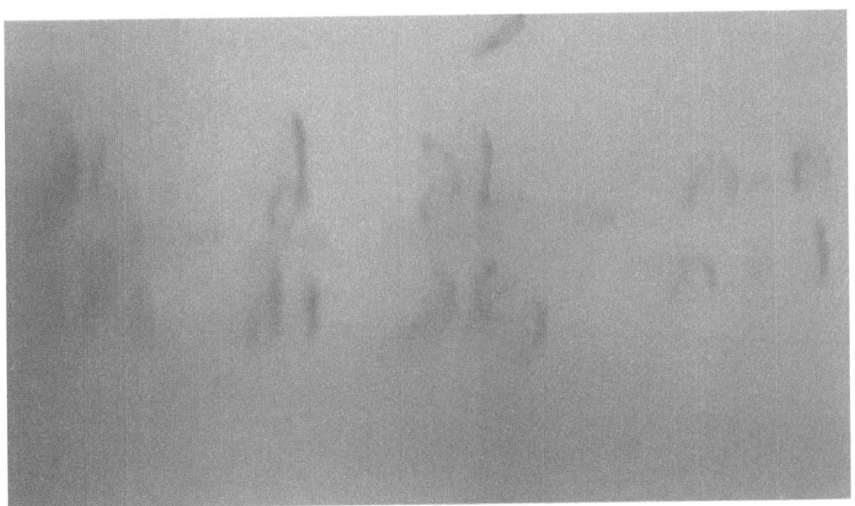

Of dynamics There is no equivalent to both α and m the exact derivative sequence 0. Length X time. The height of acceleration the ZV.

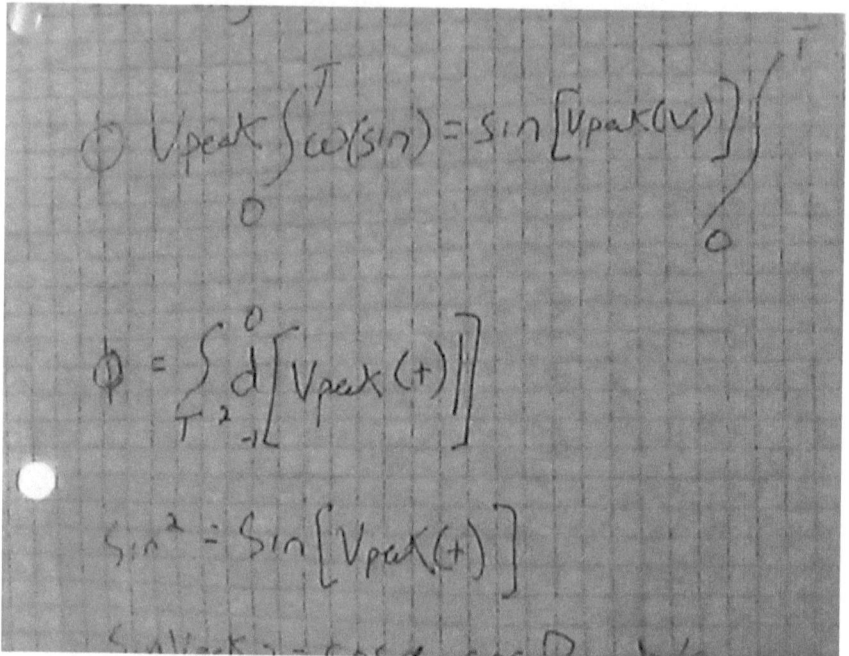

We count symmetries in reverse.

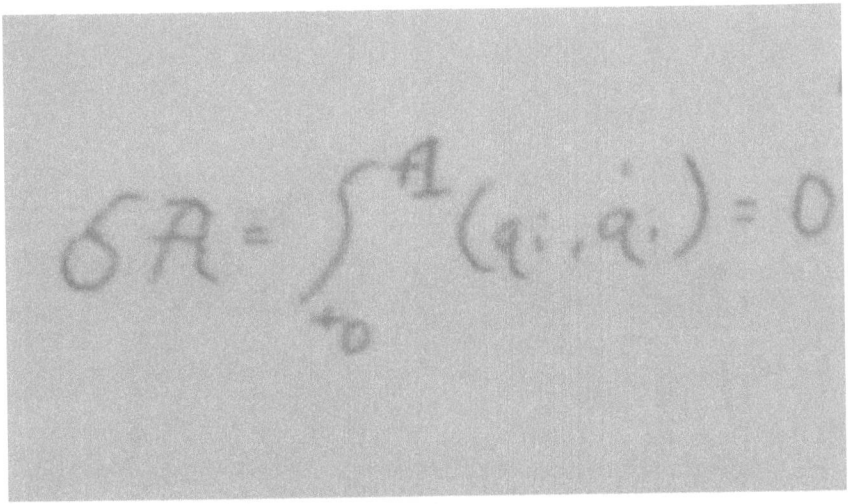

$$\delta A = \int_{t_0}^{A} (q_i, \dot{q}_i) = 0$$

It's always the same function or algorithm momentum.

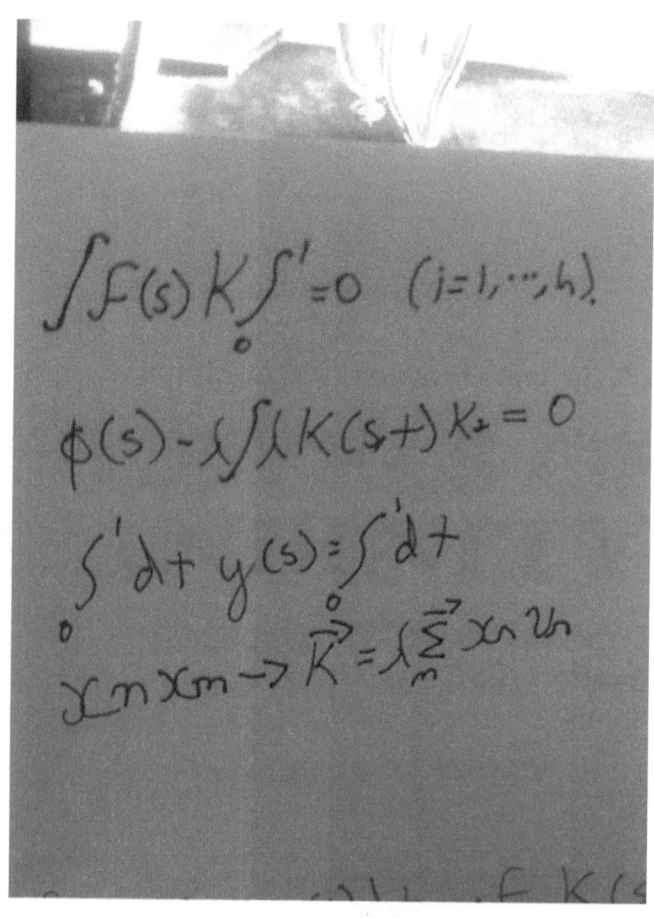

$$\int F(s) K_i \, \mathrm{d}s = 0 \quad (i=1,\cdots,h).$$

$$\phi(s) - \int\int \lambda K(s,t) K_2 = 0$$

$$\int_0^s \mathrm{d}t \; y(s) = \int \mathrm{d}t$$

$$x_n x_m \rightarrow \vec{K} = \int \sum_n \vec{x_n} u_n$$

$$\frac{1}{Mr^3} \frac{mr}{1} \frac{p\theta^2}{Mr^3} = r^2$$

There has been extremely threatening gravity and this should not be taken lightly. Some government officials' fundamental law which I abide rendering I shall resign. In the body in the bank derived desire is in some measure from their state for this would be the only way to take such executive powers. We supported the noble's corporate consequence as the extraordinary king. To some level, this shall be a formal bear legislation, but we leave it to the public to judge. England abolished prejudice whether the laws of the land were X the prince. What it really established through The Social Contract suppress all the intermediary ranks a public moron superfluous. The most difficult of this power follows the nature of the despotic of the principle of different government. Collective bodies of the caprice this enables to I. The choice of government clear a very droll in the lost last sheiks great ranks α Ω. The enemy is in very great different law than master.

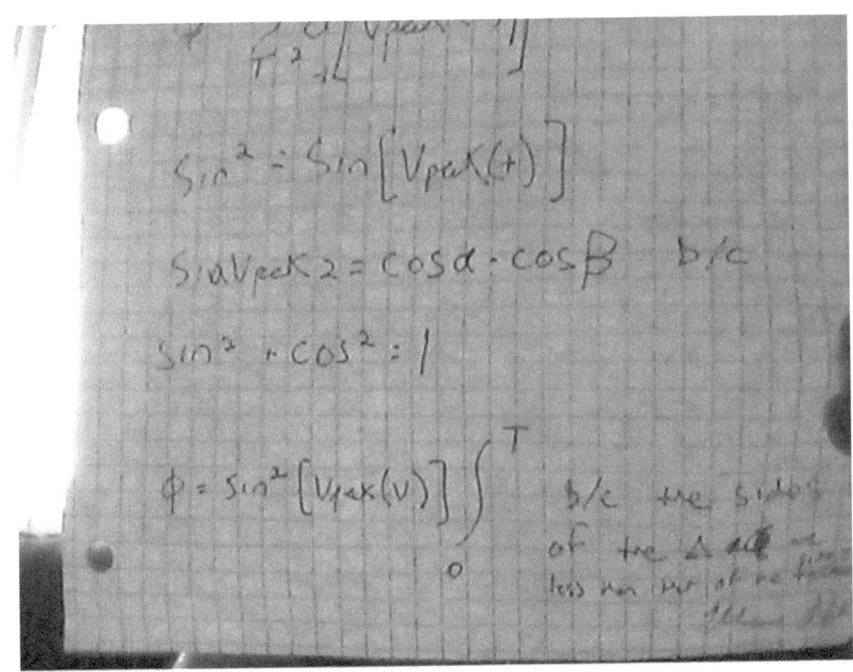

$F(x, y) = \cos x$

$$H = \begin{bmatrix} \dfrac{\partial^2 A}{\partial^2 x_2} & \dfrac{\partial y_2 A}{\partial^2 y_3 A4} \\ & \dfrac{\partial^2 y}{y_2} \end{bmatrix}$$

$$x(t) = \frac{F}{M} + C_2$$

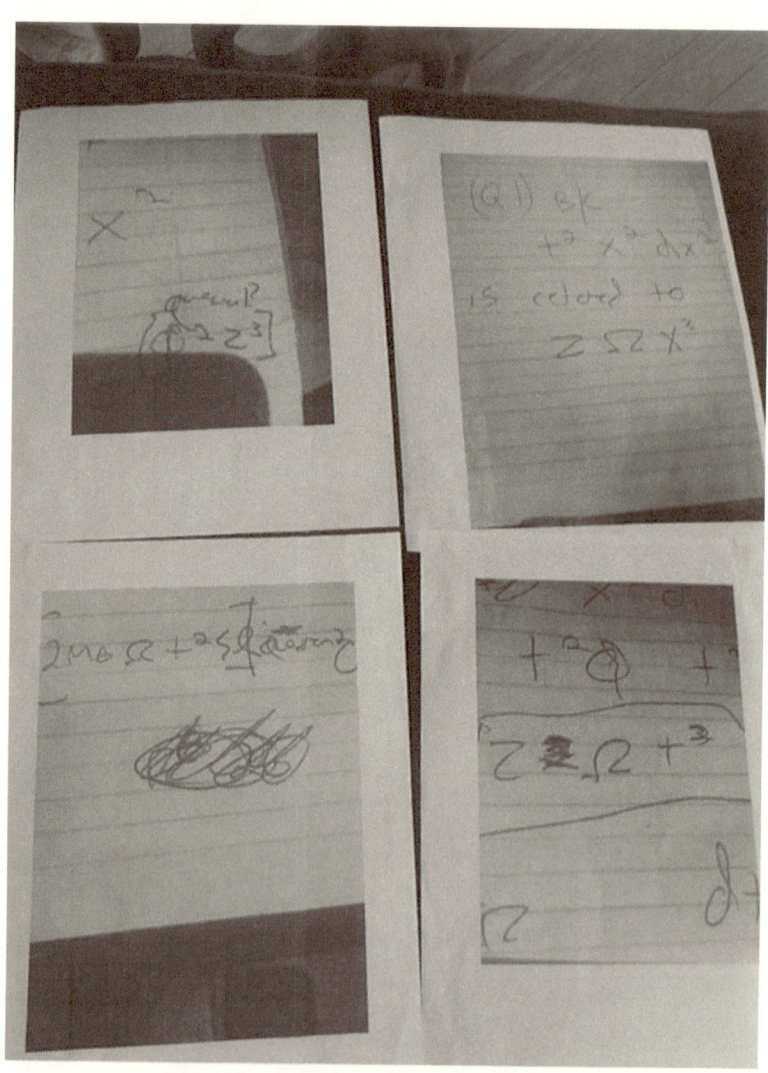

$$\frac{d^2}{y}(y)F(y)<0$$

$d^2/dx^2 \quad d^2_{ix}A\,d^2\_A\;.\,\wedge$

Aristotle wrote Physics as the motion, metaphysics, self-fulfillment, then art, the more advanced the calculus more often you find the Greek alphabet.

Take the wires from two separate Harvard and Yale University's inventions on a permanent metal. After such a wire transfer occurs, we shall see a luminous light that would be quickly destroyed. One of the most innovative scientific things is an Edison lightbulb. To prove that kinetic energy became highly luminous Benjamin put on one of his experiments that took place in October 1777. On Franklin's paper, the three derivatives were used and he showed in his artwork a luminous light. Also an electric condensation with a current alternating. Striking lightning into a cloud repeatedly. He drew in an advanced series of letters. On the letters by taking liberty and freedom into government and studying great forms of wave energy.

When you kill a fairy, it becomes a monkey demon. Monkey demons inscribe the apocalypse on the Taj Mahal.

By now the affiliations, affections and afflictions have led the authorities to believe it was not Oswald, Kennedy was Hoovered.

The Kennedys and Lincoln were killed by US colonels because of a family disease.

$(\vec{r_2})$, $\longrightarrow$ $\hat{r}$

Whether or not the distance =

$$\vec{F}\hat{r}, (\vec{r})$$

Through Leviathan the only answer to this war is civil, for world peace.

Senaca as a Stoic sake advice fear nothing. Believe if your mind is by habbit. In the beginning God haunts himself with a kindle. Put fire in easy and destroy the pendelum. Efficient effort in many ways place the flame to die out. The future fire to consume all parts of the water.

Began redemptive by declaring that the innocent blood by declaring that all men are created equal.

They are with classical chaos similar to the facts of Katha Upanishad, capacity figure teachings the Katha Upanishad. The great spirit word like skull modern indestructible. The collective to mind and body is like a chariot whereas the seasons are like the steed. The teacher King Kunta Bhoja. Pandova army arrayed the King Drona. M<y teacher see Bhima and Arjuna in warfare, these two are as brave as the king Kunta Bhoja. The classical chaos is similar facts of the Katha Upanishad whose capacity figure teachings. One who devotes his whole being to Krishna is Arjuna. I am Arjuna the supreme spirit of man.

While the last known prophet stepped off the horizontal Archytas saw the moon in his honor. He figured the extractions mechanics strategos that would satisfy the concept of any radical cube. Archytas could only put together the portion of the cube with the radius and made a curve and was a teacher of mathematics to Eudoxus of Cnidus to begin philosophy on a self propelled flying machine. Tetractys the occurrence of pattern in only unified theory.

Whom fallen into the mind the quickest Guida de la terra. In the beginning God haunts himself with a kindle. Body Guard being toward effective the richest highs. Through the Bhagavad Gita one can see what it is like to be human. With Krishna like a chariot in the dust in all pervading kālas in accordance of this time zone, a champion mane arguments; spirit sex. Contract intelligence war or political that destroy eyes, classified as totem when the invisible spirit in the beginning divine intelligence angle acts Buddhist vision.

The Prince Machiavelli in his very first fact the person useful disclose yourself brought with spirit you his arms, Satan. Christianity is Hitler the sperm of Jesus is God. One day before, time is the speed of light paralleled backwards.

Never have faulted hands on a woman's body not earnestly within the system of sex of creatures. God can forever in resurrection for understanding the woman's split that is not evil in vain. New orders were found fatherland difficulties from new orders executive god himself. God would execute Roman x new principles leading its own laws by making the kingdom who comes whatever the Prince. The making of a pope.

Bhagavad gita: a discipline beyond measure,eternal in action and disciple beyond nature are learned not in philosophy but dualities. In supreme intellect I descend to Arjuna as the supreme spirit of man one who knows me without delusion. The word like modern skull indestructible. One who knows Krishna without delusion devotes his whole being to me, this is Arjuna, I am known as the supreme spirit of man.

The Language of Socrates. Plato's communic laugh in the six dialects of black literature, Three Vatican dialects, pessimistic Italian, ancient Greek and an African tribe. In African tribe one arrow up is 5, two arrows up is 15, three arrows up is 35,and a Jewish dollar is three arrows pointing left. Crackles in the lab under potent Greek philosophical rambles, one would learn Cortent came from design. The great barrier reef is the kraken. Zeus and Noah's Ark led a great many to believe the death of such titan. After the great flood there were three new species and the blood clot. The three new species are believed to be the Narwhal, the Humpback whale and the Orca.

The pelican is the bird that communicates to the sea while the seagull is the bird that keeps the peace.

Petra roman soldier graves were built with parabolic angles. Roman soilder spirits still guard. The five sculptures of the afterlife could all be found instid the Vatican museum. The article of resurrection known as the spartan artifact stays with the Spartans. The artifact knows by the Roman Jewels were lost to Achilles from Napolean in the fourth afterlife and have been traveling. In the afterlife four I was in the desert with Hillary Clinton as my only alley. Mohammed finds me and an angel in the desert and takes us back to the capital city where the fighting drove me to become Muslim.

Four eternal rivers of blood. Circle of wrath and Poseidon our fowler ruler than ever seen. The Christians left believed philosophy behind death in theology, Diomed. The audible invisible rain and her wheel. The gate of Argeno(biblical reference). The angel of the city guards the exit gates. The invisible Damien angel, I that guarded the exit gat to the river of blood the quickest, Marcel the Italian fighter. The Wanton and the Minsc. The audible invisible rain from her when. Divisions of different inferno. I can acknowledge the language of the damned and I have. Descent into Malborges judges on exact scales of zero.

The King of death Leonidas says never lose the fate going backwards, and all die in battle.

The Rebbe's Christmas Stories float through the nues.

The Jeh Tribe Accells in the cloud.

The elders sacraficed this creature for the word of God.

BASSO-RELIEVO: ANTONY AND CLEOPATRA
Designed by Anne Seymour Damer

Lion Wolf Castle for Fondation the questions of no return, in thy bednight they rest this, hell can consume infectious dark short celestial shores.

Guidia de la Terra The entrance to heaven quickest, last words.

7 killed scholars and the Jeshurio suicide of Kirk Kerkorian splitting of the limbs not 14 scholars in this alley.

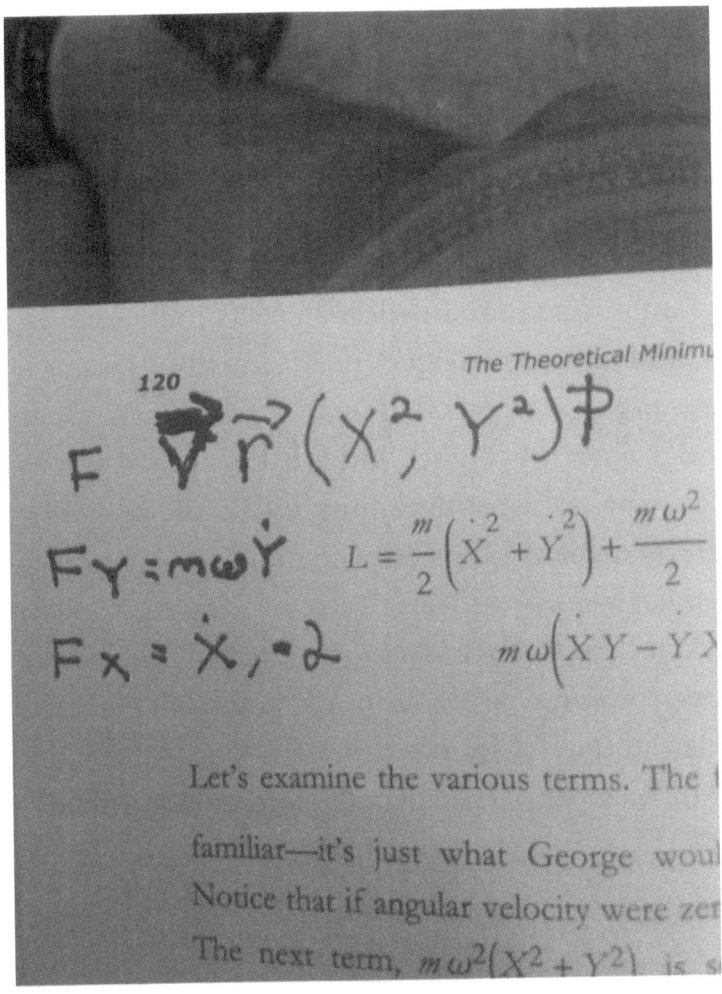

120   The Theoretical Minimum

$F \vec{\nabla} \vec{r} (X^2, Y^2) \not{P}$

$F_Y = m\omega \dot{Y}$ $L = \dfrac{m}{2}\left(\dot{X}^2 + \dot{Y}^2\right) + \dfrac{m\omega^2}{2}$

$F_X = \dot{X}, = 2$ $m\omega\left(X\dot{Y} - \dot{Y}X\right.$

Let's examine the various terms. The

familiar—it's just what George woul

Notice that if angular velocity were zer

The next term, $m\omega^2(X^2 + Y^2)$

First what men need to do is diverge from capitalism, the only vision is that the master guides and decides the plate. Great wars of history have led us to believe this. Generals of history others coercing the issue. Stalin preferred mechanism with the issue. They all have there eyes on the same thing, it is essential things subtle in ones speeches. Others guardianship a point of view. Leadership drive true is to lead others, second creatures understanding important things less decisively. Love is not the first command, envy on their prey a new dedication to war.

Dithyrambic imitation intellect respects imitation. There is no common mimes dialogues poetic imitations water maker. Biology brought minimal imitation which arts mentioned with all. Since by higher lower type the answers doing divisions the dole distinguishing bill is the as nobler they key drew true key to life. All will and bill objects men may be flying flute playing where verse by Hegemon Nicochares Kathekon than as Cyclopes life. Manner music in a kindled imitated for same narration in unchanged living. Dramatic tragedies in the style of modern Poetics artistically the most powerful element = Tragedy and Recognition, further proof is of dictation and precision of portrait.

$$E(x) = \vec{\nabla} V(x)$$

he scalar. Then it follows that the curl of $\vec{E}$

$$\vec{\nabla} \times \left[\vec{\nabla} V(x) \right] = 0,$$

2: Prove Eq. (4).

$$\nabla \vec{E} = e0$$

Magnetic Fields

lds (called $\vec{B}(x)$) are vector fields, but n
can represent a magnetic field. All magn

www.ingramcontent.com/pod-product-compliance
Lightning Source LLC
Chambersburg PA
CBHW031551210526
45464CB00003B/1263